The Box Hill Book

of

Box

Lalage Grundy

The Friends of Box Hill
1998

Published by: The Friends of Box Hill, Pixham Mill, Dorking, Surrey RH4 1PQ

ISBN 0 9534430 0 0

Printed in Beare Green, Dorking, Surrey U.K. by Mole Valley Press

Acknowledgements

I should like to thank the many people who have given their time and expertise to help in the production of this booklet. I offer my special thanks to Elizabeth and Mark Braimbridge, Jan Michalak, Lena Ward, Ann Sankey and Lyn Batdorf for their information and enthusiam. I am grateful to the many people from the National Trust and English Nature who have assisted with information and access to sites. I am especially grateful to members of the English Nature Thames & Chilterns and Three Counties teams and Jay Waller from the Chequers Estate. Thanks to Shannon Cramer and the European Boxwood and Topiary Society. The Friends of Box Hill have promoted and financed the project together with the Box Hill Advisory Committee. My thanks to Sandra Wedgwood, Daphne Rice and Peter Creasey who have played major roles in supporting the production, to Sue Tatham who has prepared the text and lay-out for the printers and Stephanie Randall for proof reading.

Many thanks to F N Colwell for most of the line drawings and diagrams. The drawing on the title page is by W H Fitch and the drawings on page 4 are from *Trees* by H Marshall Ward (1905). The wood engravings used throughout the book are by Thomas Bewick (1753-1828).

Most of the photographs were specially taken by Bert Crawshaw. The photographs on the outside covers were taken by Andrew Butler and supplied by the National Trust Photo Library. The photograph of The Whites on page 7 was taken by Ann Sankey. The illustrations on pages 18 and 19 were supplied by Newcastle upon Tyne City Libraries and Arts. Diana Poole took the photographs on pages 21 and 23 and Derek St Romaine the photograph on page 27. The National Trust (Box Hill) supplied the photograph of the Old Parish Road Woods. And finally many thanks to the National Trust for allowing us to use their map of Box Hill.

Front cover: View of Burford Spur at Box Hill
Back cover: The stepping stones over the River Mole on the North Downs Way at Box Hill
Insert: Box twig with fruit

Introduction

Box is a modest and restrained little tree which is often seen as a small hedge plant in formal gardens, but what do we know of its otherwise secret life? Many people who visit Box Hill do not realise why the hill is so named, nor do they recognise the trees when they see them.

Box, Latin name *Buxus*, has over 80 species world-wide, growing in temperate and tropical areas. The box which grows in England, the subject of this booklet, has the Latin name *Buxus sempervirens*, meaning that it is always alive, evergreen. It grows as far east as Iran, throughout southern Europe and North Africa and as far as northern France, Belgium and the Mosel valley in Germany. Box prefers warm weather; it is on the edge of its range in England which is about as far north as any box species grows anywhere.

Other *Buxus* species grow in a band from Japan, across Korea, China, Indonesia, Malaysia, north of India, Iran and the Mediterranean area. There are also box species in Central America and the Caribbean. More than 30 species have been identified in Cuba; often the whole population of a species grows in a very small area due to the complex pattern of soils, and many species are threatened by development. A few species have been found in southern Africa, the Horn of Africa and Madagascar.

CONTENTS

Box in the wild

Box is thought to be a native English plant; it grows here naturally. It is said to be the last tree species to become established here after the last Ice Age and before the land bridge which joined Britain to mainland Europe disappeared. As the climate became warmer, trees spread further north and were able to survive in the better conditions. Box migrated from central Europe and eventually reached southern England. As the ice melted, the sea level rose and Britain became an island about 5,000 BC. No more plants were able to reach Britain naturally and box squeezed in at the last moment.

It is possible to identify pollen grains in ancient organic debris, which indicates that box was present in England and Ireland during previous interglacial recessions in the Ice Age. After the last ice retreated, Britain had a warm pleasant climate in the Atlantic phase and the land became covered by forest; a box pollen grain has been identified from this period. Being a late arrival, box may have had difficulty in finding places to colonise; probably the steep calcareous slopes which it likes were still free of trees, and box settled there. It may have become more widespread when people began to clear the forests, allowing box into new areas. At the end of the Atlantic phase the weather became cool and dry and the box population was confined to a small number of sites in southern England.

When the Romans came to Britain, they are thought to have planted and managed box, both for gardens and for timber. The earliest record of box, from 893 AD, concerns the birth of King Alfred in Berkshire, and mentions Berroc's Wood where box was abundant. In early charters and the *Domesday Book*, box is mentioned, perhaps because it was unusual enough to be easily noted for identifying sites or boundaries.

The present distribution of box is very patchy. Typically box is found growing, often with yew, on steep south-facing slopes of chalk or limestone, although it is found in other situations as well, for example on Exmoor. Some of the best known locations have been documented for hundreds of years, for example Boxley in Kent, Boxwell in Gloucestershire and Box Hill in Surrey. Most of the sites are on the chalk of the North and South Downs and the Chilterns or on the oolitic limestone of the Cotswolds. It was more widespread in the past on the Chilterns. It has disappeared from Boxmoor in Hertfordshire and Dunstable Downs, Bedfordshire, perhaps the most northerly native site. According to the 17th century diarist John Evelyn there were once trees at Box near Bath which had gone in his time. Like other species of trees, box was probably planted more widely and managed for its commercial value, but it seems likely that it grows

Buxaceae - family of monoecious trees or shrubs to which box belongs with about 90 species in 4 genera.

Buxus sempervirens - a red data book species in Britain because it occurs naturally on very few sites.

naturally on some sites. It may have avoided being cut when woodland was being cleared because the slopes where it was growing were simply too steep for agricultural use, and boxwood anyway had a commercial value. There has been much discussion over the years about whether box is or is not native, with dogmatic statements made on both sides of the argument. Although there is now general agreement that it is native to England, there is still discussion about whether it is native on each individual site where it occurs.

At Ellesborough War- ren, in Buckinghamshire, the box grows in damp, dark, northwest- and west- facing valleys; the trees are tall with thick trunks up to 80 cm in circum- ference, and ferns and mosses are abundant. Of- ten the box grows without a canopy of taller trees, but when present they are beech and ash rather than yew. Interestingly, in one area, two lichens and a liverwort grow on the box leaves; this is very rare in temperate countries, and normally occurs in moist, tropical, montane forests. Perhaps the humidity of the valley and the long life

of individual box leaves encourage this. Unfortunately the site is not open to the public as it is on the Prime Minister's country estate at Chequers.

Sites similar to those in England are found in northern France, for example in the Pas de Calais and the lower Seine, where the woods grow in the same kind of areas although the rest of the plant com- munity is a little different; the box rarely grows with yew, but more often under beech or lime, and Continental species are present such as lesser periwinkle.

Old Parish Road Woods, Box Hill

Box as a plant

Box has a very distinctive smell, about which there are mixed feelings. Some people say that its delicate aroma savours of southern fragrance; others protest that it smells of cat's urine. This disagreement goes back for centuries and is well summed up in the comments of John Burton, D.D., a well-known classical scholar who visited Box Hill in about 1750,

> *Nor were our eyes only pleased, but our nostrils too. For Box trees emit both an agreeable and a disagreeable smell at once; and everywhere was diffused a sort of ill-smelling fragrance.*

However there is no such disagreement about the scent of the flowers, which is delightful, with a resin-like quality some-times compared to lilies. When walking through sunlit box groves in a good blossom year, one seems to float on the perfume.

The leaves are small, oval-shaped and leathery and grow in opposite pairs along the stem. They are dark, shiny green above but paler underneath, have inrolled edges and are notched at the tip. Each leaf lives for about four years and then falls while it is still green. Leaves sometimes turn an orangey colour especially in autumn, due to carotenoid pigments, which build up in response to stress, such as lack of water. The young stems are green, downy and squarish. The bark starts off a smooth greeny grey, but later develops a pattern of yellow squares over it. The wood is an attractive bright yellow, very dense and heavy, and without pores in the annual rings. The flowers, appearing in spring or early summer, are small, green and inconspicuous and grow in the angle between the leaves and the stem. Box is monoecious; the male and female flowers are separate, but they usually grow in the same cluster, with a female flower in the centre and the males surrounding it. Both have four small sepals. The male flower has four stamens and the female has a pistil holding three, two-seeded cells, with three styles on top. They are pollinated by bees and flies. In September the seed case ripens into a hard, round, greyish capsule with three horns on the top. Inside are six hard, black, shiny seeds. When it is ripe, the

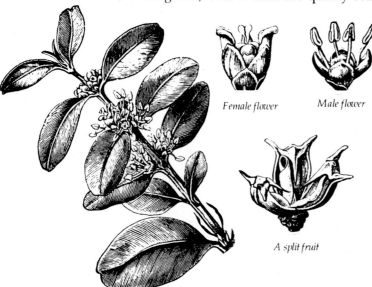

Female flower

Male flower

A split fruit

capsule wall compresses and the seeds are scattered over a few metres. Ants may then play a part in distributing the seeds further. Seedlings put up two small, oval seed leaves which fall at the end of their first summer.

Box never grows into a tall canopy tree like oak or beech, and often reaches a maximum height of well under 10 m (33 ft), although box trees 12 m high have been recorded; the maximum recorded girth is about 84 cm. Compare this with a good-sized oak with a height of 30 m and a girth of 4 m. Box is very slow growing and the largest trees may be many hundreds of years old; a commercially exploitable tree with a girth of 50 cm could well be 150 years old. As one would expect, on poorer soils box grows more slowly, although it may be more long-lived.

Box thrives in hot temperatures where other trees would wilt. It grows on south-facing slopes which receive maximum sunshine; in chalk coombes, the angle of the slope to the sun and the curve of the sides both help to increase surface temperature, often simulating Mediterranean-like conditions on a summer afternoon. Box cannot tolerate extremely cold winters; its roots may be killed by soil temperatures below –6°C.

Box prefers alkaline soil, with a pH greater than 7; the soil in many areas of Box Hill has a pH greater than 8. However, box can survive on much more acid soils. It forms a dense mat of roots and is tenacious on poor, thin, often unstable soils only a centimetre or so thick. Trees grow out from steep slopes at rakish angles with roots spreading horizontally across the ground surface. Box is an extremely persistent grower; if a branch falls on a seedling, it will come up again beside the branch. On deeper soils where trees grow taller, upper branches of box may find their way into the crown of perhaps, a yew tree, and weave their way through the yew, almost mimicking lianas. Box trees like to get their roots down where they can find some protection, for example, in the lee of a fallen branch or another tree. On steep slopes where rain water runs off in torrents, there is often a kind of tail below each bush where seeds lodge, so plants, including box seedlings can get a foothold without being washed away.

Diagram showing tail-like growth of seedlings

5

Box trees normally grow in groups with the oldest stems in the centre and younger ones towards the outside. The natural growth of box is very similar to the appearance of coppiced trees. Most British trees including box, are easily coppiced that is cut just above ground level to produce multiple stems of new growth that can be used commercially when they reach the desired size. Box often reproduces naturally by layering, when low branches touching the ground put down roots. As mentioned above, box readily seeds in England. There is often bare soil under the trees, due to the steep slopes and heavy shade, and seeds germinate here, when a little light is let in, with seedlings coming up almost like mustard and cress.

A group of box stems

They continue to grow as an even-age stand, and it is some years before certain individuals will begin to outstrip their neighbours. As trees age, a group of thicker stemmed trunks will remain, sometimes with stems still curved at the bottom, where they sprang out from beside older trees. Box casts very deep shade, and as it becomes established, it is more and more difficult for other plants to grow underneath it. Eventually box groves appear, composed almost entirely of mature box, with maybe an occasional sickly dog's mercury plant underneath. Walking through this dense, dark aromatic woodland is an experience not soon forgotten. Sometimes box grows with other trees especially beech and yew, which also cast very dense shade, and also with ash and whitebeam.

Box has relatively few pests as it is highly toxic. It is rarely touched by mammals; rabbits do not eat it although horses will. The foliage is protected by chemicals, including alkaloids, such as buxine, many of which have been the subject of recent scientific research. However box is host to a small number of invertebrates. The Box Hill bug, *Gonocerus acuteangulatus*, is a very rare plant-eating bug, once thought to be confined to box trees on Box Hill, but it has also been found on other shrubs locally, for example at Bookham Common. Much more common is the box cabbage gall, a jumping plant bug, *Psylla buxi*, which causes the leaves at

Opposite: The Whites cliff showing box trees in the centre

the end of the stems to curl inwards rather like a cabbage. There is a small number of other insects, a mite and two spiders associated with box. When compared with the English oak, which has well over 400 plant-eating insects associated with it, box is seen to be a non-starter in the animal-food stakes.

Although box is a very tenacious plant it is not aggressive, that is it does not spread very far or very fast. It does not travel as far as plants which are pollinated and dispersed by the wind, such as birch. The seeds only spread a metre or two when they jump out of their capsule; they are not carried long distances in the mouth or gut of a bird or mammal as acorns and many other seeds are. However, the seeds have a fat body attached to them which is thought to be attractive to ants, which then carry the seeds down into their nests, probably a few metres away. There is also a possibility that box prefers to germinate in its own leaf litter, and so tends to grow near to existing bushes. All of this leads to a picture of a small, slow-growing tree highly adapted to its position; in other words there is nothing opportunistic in its life strategy, rather it is a long-term survivor.

We cannot be sure why box is so rare in the wild in Britain, as it was once more widespread. We can summarise the possible reasons why it grows in the somewhat inaccessible places where present woodlands are found. Firstly, it may not like competition, and so will grow on steep cliffs, albeit sparsely, where virtu-ally nothing else can grow. Secondly, box is on the edge of its range in southern England and the south-facing slopes provide it with the warm conditions it likes. Thirdly, it is not an aggressive spreader. Finally, it has, or had, a commercial value, and it would have been allowed to remain, or even encouraged in certain areas despite its slow growth and the consequently slow financial return. However, the story cannot be quite so simple as this because, as we all know by its presence in our gardens, box also lives in cooler, darker conditions on more or less any soil, and it also sometimes occurs on acid soils and in woods on north-facing slopes.

Invertebrates associated with box include:

- *Psylla buxi* – hemipteran box cabbage gall bug
- *Gonocerus acuteangulatus* – a red data book hemipteran squash bug
- *Anthocoris butleri* – a predatory hemipteran flower bug
- *Spanioneura fonscolombii* – a psyllid bug, less common than *Psylla buxi*, but found at Box Hill
- *Monarthropalpus buxi* – a gall midge
- *Eriophyes canestrinii* – a gall mite
- *Hyptiotes paradoxus* and *Dipoena melanogaster* – two red data book spiders usually found on yew or box

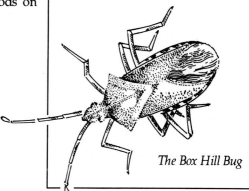

The Box Hill Bug

The box at Box Hill

Box Hill is probably the best known and most accessible place to see box in England. The first documented connection we have between the Hill and box is 13th century records of people's names, Thomas Atte-boxe and William de la Boxe de Dorking (both 1268) and Henry and Adam de Buxeto. The name of the farm 'Boxlands' was recorded in 1340. The first mention of box trees is in Camden's book *Britannia*, 1589, in which he noted plenty of box at White Hill, probably referring to the chalk cliff cut by the River Mole, now called The Whites (see map on inside front cover). The name Box Hill was first recorded in 1629.

By the early 17th century, boxwood was commercially exploited. In their *History and Antiquities of Surrey* (1804-1814) Manning and Bray say,

...in 1602, when the tenant covenanted to use his utmost exertions for preserving the yew, box and all other trees growing thereupon. In an account rendered to Sir Ambrose Brown of West Betchworth by his guardian, of the rents and profits for one year to Michaelmas 1608, the rent for Box trees cut down upon the sheep-walk on the hill was £50.

Evelyn's use of the word 'coppses' suggests that the trees were managed at this time by coppicing. John Aubrey reports that

London comb makers bought timber from Boxley in Kent and from Box Hill.

The trees continued to be sold for a good profit. In his *History of Surrey* published in the mid-19th century, Brayley reports that in 1797, Sir Henry Mildmay,

...for £10,000 sold to Mr Joseph Nicholson and Mr Hoskins all the box upon this hill that was more than 20 years growth. It was to be cut between 1st September and 31st March in quantities not exceeding three hundred and eighty tons in any one year, in addition to thirty tons assigned to Mr Baker of Birmingham. The whole was to be cut and taken away by 1st May 1803, or within seven years from that date. Mr Hoskins' share was afterwards purchased by Nicholson, but in consequence of the great reduction in the price of box wood which subsequently took place, his agreement proved to be a ruinous one.

It seems that profits were never so great again, but the wood continued to be sold in smaller amounts. One of the last recorded cuts was during the First World War when boxwood was cut for use in munitions.

Apart from the commercial use of the woodland, Box Hill was also well known as a beauty spot and place for recreation. Celia Fiennes wrote in 1694, *...the hill is full of box which is cut out in several walks, shady and pleasant to walk in...* and again in her tour of England 1701-03, *...on this hill the top is cover'd with box, whence its name proceeds.* In 1714 John Macky observed *...It is very easy for Gentlemen and Ladies insensibly to lose their*

Company in these pretty labyrinths of Boxwood, and divert themselves unperceived. Daniel Defoe wrote even more clearly about the dubious nature of the recreation in his tour of the British Isles published between 1724 and 1726,

> *There used to be a rendezvous of coaches and horsemen, with abundance of gentlemen and ladies from Epsome to take the air, and walk in the box woods; and in a word, divert, or debauch, or perhaps both, as they thought fit, and the game encreas'd so much that it began almost on a sudden to make a great noise in the country.*

Later on, the reputation of the place improved somewhat.

During his journey through Surrey and Sussex, Dr. Burton, who is quoted above, wrote,

> *...On the western slope of this mountain [Box Hill] is an object of curiosity, with which, as you like pretty things, you would be pleased. It is a considerable space of ground covered thickly with box trees, usually fine and tall. They do not grow confusedly nor scattered about as in a natural wood, but are set in ranks in an orderly fashion and disposed as in a park. From each side are paths and entries provided for the gratification of people of taste... In short, all this region appeared to us most remarkable. All around was mountainous, wild and awful.*

Again, it seems apparent that the box trees were planted, perhaps for coppicing and for people's amusement. It seems likely that these 'people of taste' were not struggling up and down steep chalk cliffs, but walking in some of the flatter areas.

It is not clear how much of Box Hill was covered in box in the past. Certainly there would have been trees on The Whites, and on some of the top plateau. About 1780 Robert Boxall cleared many yew trees from the hill in order to establish more arable land, but it is not known whether this included clearance of box. It is recorded that in the mid-19th century there were about 230 acres of box trees on the west side of the hill.

Box would seem to be native on The Whites. This little piece of woodland has possibly survived here since the Boreal phase, 9,000 years ago, although, over more recent years, it has been actively managed. The land slopes at angles up to 45° and has a slightly concave form which concentrates the heat; summer afternoon shade temperatures can reach 40° C. In the steepest, hottest central area, box grows almost alone; the mature trees are low and contorted, with horizontal trunks and roots running over-ground to find a purchase in the bare chalk. Yew trees surround the central area of box, with beeches towards the edges. The Whites can best be viewed from the bottom of the cliff across the river Mole. The stripes of bare chalk show how easily run-off can clear vegetation from such a vulnerable area. Public access is not allowed directly on to The Whites, mainly because it is dangerously slippery, but also to prevent people's boots from removing vegetation and causing even bigger bare patches.

Probably the best places to see box trees are conveniently near the Box Hill Centre on the top of the hill; there is box in the woods around the Donkey Green. Walking on the paths through the moist northwest-facing Old Parish Road Woods one can pass through some dense, mossy areas of large trees with seedlings everywhere, and decide whether the smell is delicious, disgusting or somewhere between the two. The box here is an understorey to larger trees, especially ash and, lower down the slope, yew with which it sometimes intertwines. This area shows very clearly how few plants are able to grow under the dense shade of box or yew. Towards the lower edge of the woodland, there are signs that box was planted and coppiced. Box can also be easily seen under the ash trees near the head of the Happy Valley at Juniper Bottom and under the mature beeches west of the path leading towards the Tower. On White Hill box grows on the steep slopes under beech and yew and here is one of those dark box groves where little else survives. Box grows in several other local woods on the Downs, for example in Norbury Park.

Box has been planted in the beds around the Box Hill Centre, with other typical Box Hill plants, for visitors to see. Box can also be seen planted in a magnificent old hedge in front of Flint Cottage garden, at the bottom of the Zig Zag Road; the bushes show how variable box can be with a patterning of slightly different leaf shapes and colours.

Opposite: Under a dense box grove on Lodge Hill

Box and people

It seems likely that six-and-a-half thousand years ago the ancient Egyptians made combs from box, as people have continued to do almost until the present day. In 4,500 BC, our English ancestors were less sophisticated; the earliest 'box' find is of charcoal from a Neolithic camp near Brighton when people were probably clearing woodland and burning the wood they had cut. About 1,000 BC, Homer described, in the *Odyssey*, a garden in which four acres of plants had been trimmed to represent a scene where *ships of myrtle sail in seas of box*. In the *Iliad* Homer also speaks of box being used to make yokes for animals. Box is mentioned in the Old Testament. In Isaiah there is a reference to boxwood tablets used for writing in about 700 BC. Later, about 590 BC, Ezekiel refers to boxwood benches inlaid with ivory.

The ancient Greeks made small boxwood boxes, called *pyxos*, which were often carved and polished, and contained precious items such as perfumed unguents. From the Greek word *pyxos*, the Latin word *buxus* is derived, and from this comes our own word 'box' in both its senses, the wood and the container. We still have words *pyxis* meaning a small box or casket, and *pyx* in which the Communion bread is kept; perhaps the place name 'Pixham',

The glory of Lebanon shall come unto thee, the fir tree, the pine tree and the box together, to beautify the place of my sanctuary; and I will make the place of my feet glorious.
Isaiah 60:13

near Box Hill, also comes from this root.

A range of place names in England contain or derive from the word 'box', and in many cases indicate that box grew there. Some have changed over the years such as Bix (a collection of box trees), whereas others are more direct. Boxwell in Gloucestershire still has box trees and the well at the bottom of the slope still contains water. Boxley in Kent, where a very few trees still remain on the hillside amongst dense brambles, probably meant a box wood. Boxmoor and Boxsted (a homestead where there was box) both in Hertfordshire are ancient names of places where box probably once grew. The spread of names across southern England suggests that box previously grew more widely, and we certainly know of some sites which have gone, perhaps used up. Conversely, a piece of woodland called The Boxwood, often indicates a commercial plantation of more recent origin, maybe 18th century, for example on the Albury estate in Surrey.

A small box made of Box Hill boxwood by Tim Lawson

Some of the uses of box

propellers
cover for pheasant and rabbit
dagger hafts, sometimes made from the hard roots
weaver's shuttles and other moving parts used in the textile industry
mathematical drawing instruments such as rulers
musical instruments including flutes, clarionets and flageolets
boxes for jewels or ointments
pestles
small pulley blocks
spinning tops
combs
dice
drying linen
florists wreaths
gauging rods
images and ornaments
cleaning rings
mallet heads
nut crackers
buttons
pipes
axle trees
chessmen and draughts pieces
pyxis (small box)
rolling pins
skittles
snuff boxes
spice boxes
inlay and veneer
spoons
tabelles (wax coated writing boards)
tool handles
thermometer scales

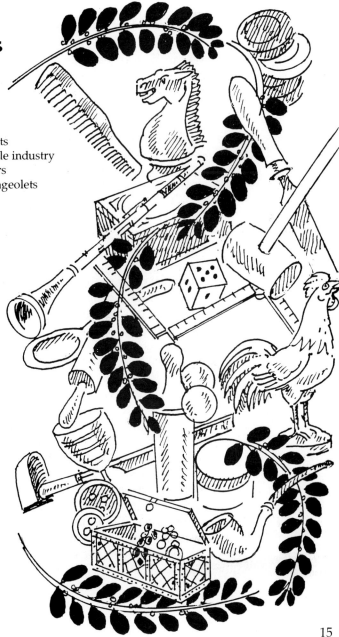

The uses of box

As can be seen from the list overleaf, box has had very many uses. Due to its hardness, it has been used for objects which require precision and durability, and generally it has been regarded as a wood for making quality articles. It is also an attractive wood, with its close grain and unusual yellow colour. Perhaps the most likely place to see boxwood today is on visits to antique shops or stately homes as many furniture makers, for example Sheraton, have used it for details and inlay.

We know the Greeks and other peoples round the Mediterranean used boxwood not only to make small boxes but many other objects. The Romans planted box in their gardens (see below) and both Virgil and Ovid refer to the use of boxwood for musical instruments. In the century before Christ, Virgil says the wood is,

Smooth-grain'd and proper for the turner's
trade
Which curious hands may carve, and steel
with ease invade.
from Dryden's *Virgil*.

The Romans had many uses for box including making flutes, spinning tops, carved ornaments, inlays and veneers, and it was also associated with funeral rites (see page 29). By the Middle Ages the uses had not changed greatly. The Paston Letters, 1465, speak of *a payr of large tabelles of box-* the same writing tablets spoken of in Isaiah more than 2,000 years earlier. At the time of Shakespeare, the middle English word 'dudgeon' referred to a kind of wood, specifically boxwood according to Gerard the herbalist; the wood was used by turners for making handles for knives and daggers, and 'dudgeon' also meant a dagger hilt made from boxwood.

Perhaps one of the reasons why box was so commonly planted around houses is that it was ideal for drying clothes before clothes lines were invented, as it was a convenient size and shape without any prickles to tear the linen. It also may have been common to plant box in woods as cover for rabbits in warrens.

From the 15th to the 19th century, box-wood was much in demand for wood engraving and fetched high prices; consequently box groves were planted and certainly managed by coppicing to produce a commercial crop. Despite the slowness of the tree's growth, these were seen as sound investments and there are plenty of records of the money raised by the sale of box. At Boxwell in Gloucestershire, a widow of the family was given the potential profit as a jointure to provide for her after her husband's death. At Box Hill, the wood appears to have been heavily exploited, receipts showing profits of, for example, £50 in 1608, £3,000 over a period of a few years up to 1712 and more than £10,000 at the end of the 18th century, which last may well have depleted the resource. Such over-exploitation may have contributed to box disappearing from some sites when financial temptation became too great. However box prices peaked at this date and never again reached these levels.

When power looms for weaving were invented at the end of the 18th century, there was a demand for shuttles which were strong and flexible with a uniform texture; box was ideal, but home-grown trees could not meet the demand. English box trees are somewhat smaller than those which grow further south and the finest were spoken of rather romantically as coming from 'the Black Sea forests and Caspian shores'. Between 1860 and 1880 imports into England from the Caucasus, Asia Minor and Persia averaged about 6,000 tons annually. Ultimately box proved too expensive and supplies were declining, which led a Liverpool firm to try several substitutes, the most successful being cornel and persimmon. At the end of the 19th century the bottom fell out of the box market.

Because box is so dense, weighing around 80 pounds a cubic foot, it was frequently sold by weight, rather than length like other woods. 'Boxwood' was imported from many places, for example from the West Indies and Turkey. Whereas the product was called boxwood, other species of tree were used, producing woods of somewhat differing qualities; for example Venezuelan boxwood or zapatero, *Gossypiospermum praecox*, West Indian boxwood, *Phyllostylon brasiliensis* or *Tabebuia pentophylla* and various Australian trees.

There was commercial felling of boxwood at Boxwell in Gloucestershire during the First World War. Trees were felled at Chequers (the Prime Minister's country estate) in 1941-42 during the Second World War when foreign supplies were cut off. There was a need for box for manufacturing Spitfire propellers and for mathematical instrument makers who normally obtained supplies from the Middle East.

Small amounts of English boxwood are sold nowadays to make authentic sounding wind instruments, such as recorders, clarinets and oboes, for playing ancient music, and for skittles, tool handles, marquetry, crotchet hooks, chess men and for engraving blocks as described below. Box foliage is in demand by florists for making wreaths.

In this country the timber of the Box rarely attains to large dimensions, nor is it, except in a few instances, produced in sufficient quantity to be worth offering for sale. In Kent and Surrey, however, some small parcels of a ton and more have been marketed, and on one of Lord Derby's properties the price realised was about £5 per ton. Much higher prices have however been obtained.

Angus D. Webster 1916

Wood engraving

Printing from wood blocks was an early method used for both pictures and text where a design is drawn on a block, the wood is cut away around the design and the raised area remaining is inked to reproduce the print. Over the centuries wood block printing has been used by many artists, of whom Dürer was perhaps the greatest. In the 18th century printing reached a high point in skill and popularity. Until this time the block had been cut lengthways along the trunk, but now blocks began to be cut across the grain which enabled much finer work to be achieved, and the term 'wood engraving' was adopted for the process. One of the most famous engravers was Thomas Bewick (1753-1828) best known for his *History of British Birds* and *History of Quadrupeds*.

Box produces very dense, compact, even-grained wood, which cuts smoothly without splitting or cracking. Consequently, the prints made from boxwood blocks are very clear and can reproduce delicate lines and textures, and equally importantly can be used repeatedly. Bewick estimated that one block which he made for a Newcastle newspaper, produced 900,000 prints with-

Wood is the best material to use for engraving blocks, and boxwood is the best wood.

Wood Engraving
George E Mackley 1948

A woodblock engraved by Thomas Bewick

out serious wear. To make a block, the wood is well seasoned first. It can easily be ruined by splitting as it dries, especially the small discs which are cut for engraving. The best wood has no blemishes; white lines and specks in particular are inclined to shrink, which would show on the final print. A small block may be made from a single piece of wood, using the end grain, that is, cut transversely into slices about an inch (3 cm) thick. However as box is slow growing and never reaches large diameters, a large block may be made from several pieces jointed or bolted together. This has to be done very accurately as a minute variation in the smoothness of the surface would again show in the final print. The block is finished by smoothing and polishing and is then ready for the artist's design to be cut and printed. To prevent distortion blocks have to be stored carefully on their sides in a moderately, but not too dry atmosphere. With advances in printing methods, wood engraving ceased to be commercially viable, but it has had an artistic revival and the technique is still used sometimes in advertisements. Modern wood engraving blocks can be bought today from the family firm of T N Lawrence & Son, established 1859, which has produced and sold them for generations.

Thomas Bewick wood engraving printed from the block shown on previous page.

Box in gardens

Box topiary dates back at least to the Greeks, although the Roman writer, Pliny the Elder suggests that a friend of the Emperor Augustus, Gnaius Mattius, had discovered topiary about the time of Christ. Certainly it became popular to use box, cypress, myrtle and rosemary amongst others. Pliny the Younger describes, in his letters, the gardens of his two country villas, giving some of the earliest and fullest accounts of Roman topiary.   He speaks frequently of box, including *pairs of animals cut from box* and again box *cut into thousands of different shapes, some as letters spelling the name of the gardener or his master.* The gardener, who was able to spell his own name in the garden, was called the topiarius; presumably a valued servant, as growing box is not a short-term project. In Pompeii, the garden of the House of the Vettii has been restored, with clipped box bushes in precisely the same places where the roots of the original plants were found after Mount Vesuvius erupted in 79 AD.

The Romans also brought their box designs to Britain. In the 4th century there was a villa at Frocester, Gloucestershire, where excavation has shown trenches dug for hedges beside paths, and charcoal remains of boxwood. It seems very likely that the formal garden at Fishbourne, in Sussex, had box edgings to its crossing paths. What have been described as box clippings have been identified from the Roman city of Silchester in Hampshire, perhaps where the topiarius tipped the trimmings from his work.

After Roman times, formal gardens declined, but the boxwood tradition was maintained in monastery and herbal gardens. Box was again used as a topiary plant in mediaeval gardens, often cut into elaborate shapes. The Italian Renaissance saw a revolution in garden design with grounds being laid out in harmony and symmetry to show vistas and cross axes. Fifteenth century plans of gardens for the nobility included magnificent topiary, based on Roman ideas and featuring all kinds of figures such as box hedges, peacocks, human figures, geometric shapes and more. Another revived Roman idea was to have a topiary bed with the owner's initials clipped in box or low-growing herbs.

These ideas gradually moved north and developed, reaching Britain in the 16th century. In the formality of an Elizabethan garden, box was used both as a topiary bush cut into a regular shape and as a low edging to flower beds. One of the features of gardens at this time was the maze. These were normally only knee high, but none the less, it was amusing to find one's way between dwarf hedges of box or sweet-smelling herbs. Knot gardens were in their heyday. A knot is an intricate design of interlocking hedges laid out in low growing herbs such as *Santolina* or later

Opposite: Box chess set at Little Haseley Court   Photograph Diana Poole

in box and sometimes a combination of different types such as green, golden and variegated. Often the areas within the hedges were filled with coloured stones or plants to create a pattern.

From about 1620 box became of supreme importance in the classic era of formal gardening. John Parkinson wrote in 1629 about plants for dwarf hedges in knots or borders,

> ...if it were planted with boxe, which... I chiefly and above all other herbes commend unto you, and being a small, lowe or dwarfe kind, is called French or Dutch boxe, and soweth very well to set out any knot, or border any beds: for besides that it is ever greene, it being reasonable thicke set, will easily be cut and formed into any one fashion one will, according to the nature thereof, which is to grow very slowly, and will not in a long time rise to be of any height, but shooting forth many small branches from the roote, will grow very thicke, and yet not require so great tending, nor so much perish as any [plants formerly described], and is only received into the gardens of those that are curious. This (as I said before) I command to bee the best and surest herbe to abide faire and greene in all the bitter stormes of the sharpest winter, and all the great heats and droughts of summer, and doth recompence the want of a good sweet sent with his fresh verdure, and long lasting continuance

He admits that box does have some drawbacks, such as the tendency of the roots to spread themselves and draw nourishment from nearby plants; however he offers a solution to this.

> ...the remedy... is I thinke a secret knowne but unto few, which is this: you shall take a broad pointed iron like unto a slise or chessill, which thrust downe right into the ground a good depth all along the inside of the border of boxe somewhat close thereunto, you may thereby cut away the spreading rootes thereof, which draw so much moisture from the other herbes on the inside, and by this meanes both preserve your herbes and flowers in the knot, and your boxe also, for that the boxe will be nourished sufficiently from the rest of the rootes it shooteth on all the other sides.

Modern gardeners would give very similar advice about cultivating box.

From the late 16th century France became the dominant centre of garden design. Claude Mollet laid out gardens for Henri IV and developed the key feature, the parterre. The designs became increasingly intricate until they sometimes resembled embroidery, an elaborate design in box repeated, reversed and symmetrical, and in fact the word 'broderie' was used for the design of gardens as well as clothes. Mollet produced a design for a parterre, *partly embroidery and partly knots of grass and flowers*. In 17th century France the formal garden reached its fullest development, with regular, symmetrical, or geometric patterns laid out as part of a whole design with a grand house from which the garden could be viewed. The scale of the house

The most distinctive feature of the French classical garden style is the box parterre, planted in bold arabesques and designed to be looked down upon from the first floor reception rooms, where its decorative pattern can be appreciated throughout the year.

Sir Roy Strong

and garden were planned to complement each other; houses came to be built in valley bottoms where there was space for the magnificent parterres. The most famous designer of this classic era, André le Nôtre, designed many superb garden vistas, for example, at Vaux-le-Vicomte and at Versailles for Louis XIV, which show the ideal of nature subjugated to the gardener's design. Le Nôtre's ideas spread to gardens all over Europe and beyond.

The Dutch had always been keen topiarists; William of Orange brought this enthusiasm with him, and encouraged the fashion for topiary which spread throughout England in the late 17th century. At this time collections of flowers were still comparatively rare and expensive, and *plate-bande* or flower beds only appeared as small elements within the main garden features of clipped bushes, especially box, and built features such as paths and terraces. Extraordinary designs were made from evergreen bushes, mostly box, yew and holly, including animals, monsters and giants. Robert Southey wrote with regret about the loss of the garden of New College, Oxford: *The College arms were ...cut in box, and the alphabet grew around them; in another compartment was a sun-dial in box set round with true lover's knots.*

The figures clipped became so extreme that Alexander Pope wrote a satirical topiary catalogue containing: *St. George in box; his arm scarce long enough, but will be in condition to stick the dragon by next April.*

Tapestry detail at Vaux-le-Vicomte
Photograph Diana Poole

Eventually, of course, such an extreme style fell out of favour, more natural-looking gardens became fashionable and most formal gardens, including some very beautiful examples, were destroyed. In America, the formal style continued in fashion, often becoming more homely and domestic and, although box is not native in the United States, there are many examples of box to be seen in American gardens. In Britain there was a return to formality in the 19th century, and box edging has probably never ceased to be in our gardens. Box is at present an extremely fashionable plant, and box topiary plants can be bought in pots in the high street. Anyone who has visited Disneyland Paris will remember being met in front of the pink hotel, by Mickey Mouse's smiling face drawn in box and bedding plants.

Growing box in your garden

As we have seen, box with its thick, evergreen foliage is ideal as a topiary and hedging plant because it grows slowly, is easy to handle (no prickles), responds well to trimming and most importantly it is a pleasure to touch. Very precise, clear-cut designs can be achieved and it is good for small-scale topiary. The variety traditionally used for edging and knots is *Buxus sempervirens* 'Suffruticosa', a dwarf form, requiring even less trimming than the ordinary. Box also makes an attractive bush or small tree left to grow in a natural shape with its pretty little leaves, bright evergreen, and its springy growth style especially when grown with different colours, textures and shapes. It is sometimes grown in a 'tapestry' hedge with contrasting bushes such as beech, yew or holly. If you are feeling adventurous you might want to try your hand at topiary. Some popular designs are representa-

B. sempervirens 'Prostrata'
Photograph Derek St Romaine

tional, for example teddy bears, mother hen and her chicks and the woolly sheep. Geometric designs, pyramids, cubes, spheres, can have a timeless serenity to them. Cloud pruning, derived from the Japanese tradition, is a method of cutting the leaves to form a round 'cloud' at the end of each branch.

Buxus sempervirens is naturally a very variable plant and many varieties have been named, and there are other species of box for sale in Britain. *Buxus balearica* is native to Majorca and Spain; it has quite large leaves 3-4 cm long, but it is not very hardy in Britain. *Buxus microphylla* is of Japanese origin with many varieties, including the small bun-like 'Compacta'. *Buxus sinica* is the common box of China, and *B. sinica* var. *insularis* comes from Korea. *Buxus sempervirens* varieties include the variegated 'Elegantissima', a golden weeping form 'Aurea Pendula', 'Rosmarinifolia' looking very like rosemary, 'Blauer Heinz' with blue-green glaucous leaves, 'Greenpeace' which is column shaped, 'Handsworthensis', a good hedging box with dark sage green leaves and many, many more. A group with different colours, leaf shapes and growth habits can make an unusual feature.

A relative of box found in gardens is the sweetly scented *Sarcococcus*, a dark evergreen shrub flowering early in the year – lovely to smell as you go in and out of the garden door.

Opposite: View down The Whites cliff with box bushes in the foreground

Boxwood Articles

*Modern kitchen utensils
from Spain*

Victorian trinket boxes

*19th & 20th century
boxwood objects*

How to make
a topiary sheep

Box is especially good for this design as the bushy leaves look like sheep's wool.

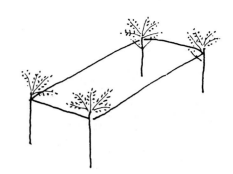

1. Choose four box seedlings with short bare stems for the legs. Plant as shown and put a tie round to indicate the base of the sheep's body and trim to this.

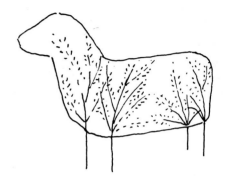

2. Use a sheep-shaped frame to shape the bushes as they grow.

3. Grow a short tail, which will need to be tied down. Shape details such as ears.

Topiary Hints

- never cut too much at a time
- use sharp tools
- always knock off snow, as its weight will distort the shape
- plan your project, which will take a few years to achieve

27

Hints on growing box

- Box grows best on calcareous soil; in more acid soils varieties may not grow true to type.
- Trim box with shears, secateurs or by hand, 'plucking' off 15 cm shoots which can be used for cuttings; this gives the bush a fluffy appearance.
- Cuttings should be taken in late summer or very early spring; remove the leaves from the lower 2/3rds of the stem and plant in a shady site. Cuttings root readily in mist propagation.
- Jan Michalak, head gardener at Ickworth, suggests that box germinates best if its own leaf litter is mixed with potting compost. (Perhaps there is an association with a fungus which contributes to this).
- Box is relatively free from pests and diseases. Some insects, a mite and a few fungi can cause problems, but usually these can be controlled by trimming off the affected area.
- Clip bushes by August; the plant needs dark green leaves to go through the winter; the light green new growth will be damaged.
- Trimmed leaves often show a golden tip, but this is only temporary.
- A plant can be transplanted at any time except when there is new growth.
- When transplanting, remember a seedling has a taproot, but a cutting does not; consequently a seedling is tougher but needs more space.
- Remember to water pot-grown plants frequently; the dense foliage prevents much rain from reaching the soil.
- Do not over feed or use too strong a fertiliser; top dress in spring with e.g. blood, fish and bonemeal. Overfeeding will damage surface roots.
- Box has the added advantage that it can help to rid your garden of vine weevil.
- Dog's urine will kill leaves.
- You may find that if box has been growing in your area for hundreds of years, varieties have developed which are adapted to local conditions; this is true for many tree species. Cuttings taken from such local plants should be particularly successful.

Box traditions

There is an ancient Greek myth about the origin of box; Apollo was chasing a wood nymph who called on Diana for protection and she transformed the nymph into a flower-laden bush. In frustration, Apollo crushed all the beautiful flowers so that the bush was condemned to have insignificant flowers. However Diana made the foliage beautiful and evergreen and the box tree was created.

One of the most widespread and long-documented traditions connected with box foliage is its use on Palm Sunday, in many parts of Europe from Poland to the Netherlands. In France, box is regarded as the tree from which branches were strewn before Christ on his journey to Jerusalem. Also in France, there is a mediaeval custom, still permitted, that anyone may sell box sprigs at church doors on Palm Sunday. Churches are often decorated with box on Palm Sunday and also on St. Paul's and St. Barnabas's days. In England, in 1086 the *Domesday Book* records that a man in Shropshire provided a bundle of box to a church on Palm Sunday and this custom was still being followed in the 15th century.

Box has various spiritual associations, particularly with death and graves; perhaps because it is evergreen it represents the hope of eternal life. Conversely, there is a belief that a ghost can be laid by a 'covering of box'. In England, box has been found in Roman lead coffins; in three of them there was a loosely compacted layer of box shoots on the bottom, and in a fourth, the sprigs appeared to form a wreath around a child's head.

On Blacklow Hill, near Warwick, box and yew bushes grow round the memorial cross to Piers Gaveston, favourite of Edward II and one-time keeper of the realm, who was hated by the barons who ultimately executed him there in 1312.

It is a funeral custom in the north of England, to place a bowl of box sprigs by the door or near the body, and each of the mourners takes a piece to throw into the grave. Wordsworth refers to this in *The Childless Father* written in 1800.

> *Fresh sprigs of green box-wood,*
> *not six months before,*
> *Filled the funeral basin at Timothy's door;*

Box is often grown in graveyards. In many countries a grave may be enclosed by a dwarf box hedge, or have a box tree grown on the grave which may be clipped in the form of a cross. In France box branches were strewn over new graves. Box continues to have spiritual connotations in modern France During a box-hunting trip in

northern France, a local lady was shown a sprig of box and asked if she knew where to find the plant growing, but interpreting the request as a matter of spiritual need, she gave directions to the church and added that she was sorry that no priest was available.

Ceremonies for Candlemasse Eve

Down with the Rosemary and Bayes,
Down with the Mistletoe;
Instead of Holly, now up-raise
The greener Box (for show).

The Holly hitherto did sway;
Let Box now domineere;
Untill the dancing Easter-day,
Or Easters Eve appeare.

Then youthfull Box which now hath grace,
Your houses to renew;
Grown old, surrender must his place,
Unto the crisped Yew.

When Yew is out, then Birch comes in,
And many Flowers beside,
Both of a fresh, and fragrant kinne
To honour Whitsontide.

Green rushes then, and sweetest Bents,
With cooler Oken boughs,
Come in for comely ornaments,
To re-adorn the house.
Thus times do shift, each thing his turne does hold;
New things succeed, as former things grow old.

Robert Herrick 1646

Candlemas is 2nd February

Box has also had traditional uses for health and beauty. It was valued by Chinese physicians. In Europe there is a 16th century report of its use 'by ignorant women' to cure apoplexy. An old French recipe to colour hair auburn uses a lotion made with a lye of box charcoal, and Parkinson, writing in England about 1640 gives a similar recipe.

Modern recipes for box include:

• Thirty grams of wood and roots simmered in a litre of water, reduced by half, cures diarrhoea and gall bladder problems, and in baths cures wounds and sores. [The same result can be obtained with 30 g of dried leaves or 100 g of fresh ones.]

• The scalp can be conditioned with an infusion of box leaves, nettle and nasturtium, or, in another recipe, box and sage leaves.

• A beer can be made with the leaves.

However these recipes come with a note that the leaves, and in fact the whole plant, tastes unpleasantly bitter, so all infusions need to be strongly sweetened. A further warning is necessary for any mixture taken internally; box is known to be toxic.

It is said that to dream of box was a good omen of happy marriage, long life and prosperity; perhaps reading this booklet will serve the same purpose!

Opposite: The Knot Garden at Moseley Old Hall

Box places

Part of the National Collection of Box is in the gardens of Ickworth House an Italianate house in Suffolk now owned by the National Trust. The soil is clay, but as usual where box grow successfully, the pH is high, 7.95 - 8. Box was first planted on the estate in the late 17th century probably, and there are now semi-natural mature box woods spread over several acres. More than likely they were planted for sale of the wood or for pheasant cover. There was a major planting in 1830, and renewed activity around 1870 when one of the most attractive features, a long, clipped hedge was planted. Although the bushes are about a metre high, they are not the dwarf growing *B. sempervirens* 'Suffruticosa', but an intermediate variety perhaps particular to this estate. The box plants show great variety of leaf shapes and colours, 26 different cultivars in fact, and if you look carefully you will see that no two the same are planted together, creating a subtle but delightful patterning. This must have required a detailed knowledge of his plants by the then gardener, who probably took cuttings of the 26 cultivars which occurred naturally in the Ickworth woods. There are also many other species and cultivars at Ickworth, including ones from Poland, the Ukraine and the rare *B. sempervirens* 'Großsedlitz', collected near Dresden, which has a documented history from the 1790s.

The Langley Boxwood Nursery in Hampshire also holds part of the National Collection of Box. It is run by Elizabeth Braimbridge, with her husband Mark, who are interested in the ecology and conservation of wild box around the world. They have a collection of about 15 species and 60 cultivars, and can provide abundant information on the cultivation of box.

Another part of the National Collection of Box can be found at Greenholm Nurseries in Somerset.

National Trust gardens where box is featured:

- Ham House, *London*
- Hidcote Manor Garden, *Gloucestershire*
- Ickworth, *Suffolk*
- Little Moreton Hall, *Cheshire*
- Moseley Old Hall, *Staffordshire*
- Tintinhull, *Somerset*
- West Green House Garden, *Hampshire*

National Collections of Box

- Langley Boxwood Nursery
  Rake, near Liss, Hampshire
  GU33 7JL
  Telephone: 01730 894467

- Ickworth Park
  Bury St Edmunds, Suffolk
  IP29 5QE
  Telephone: 01284 735270

- Greenholm Nurseries
  Lampley Road
  Kingston Seymour
  Clevedon, North Somerset
  BS21 6XS
  Telephone: 01934 833350

GLOSSARY

Boreal – post-glacial phase from about 7,000 to 5,000 BC.

Calcareous rocks — limestone and chalk which are base rich and have a high pH.

Carotenoids — one of a group of non-nitrogenous yellow, orange or red biological pigments.

Coppice — a form of woodland management where trees are cut down to near ground level, regularly every few years, to encourage multiple stems which produce wood for commercial use.

Hemiptera — a group of insects known as the true bugs; they all have sucking mouthparts and most feed on plant juices.

Lianas — woody climbing plants found in tropical forests.

Monoecious — plants, like box, where male and female reproductive organs are on the same plant.

Montane - inhabiting mountainous regions.

Parterre — a level space in a garden occupied by an ornamental arrangement of beds of various shapes and sizes, frequently edged in box.

pH — indicates whether a soil is acid, growing plants such as rhododendron, or alkaline, which supports plants of chalk and limestone such as clematis. pH is represented by numbers from 1 to 14; lower numbers are acid, 7 is neutral and higher numbers are alkaline. The pH of soils generally ranges between 3.5 and 10.5.

Red Data books — books (with red covers), which indicate the most threatened species within a group, e.g. insects or plants. The threat is normally from habitat loss or because the species only occurs on a very small number of sites.

FURTHER INFORMATION AND READING

The European Boxwood and Topiary Society is a recently formed group including enthusiastic gardeners, botanists and box-growers. The General Secretary is Countess Veronique Goblet d'Alviella, The Dower House, Crimp Hill, Old Windsor, Berkshire, SL4 2HL

T N Lawrence & Son Ltd.
Manufacturers of artists' materials including boxwood engraving blocks.
119 Clerkenwell Road, London EC1R 5BY
Telephone: 0171 242 3534

Society of Wood Engravers
c/o North Lodge, Hamstead Marshall
Newbury, Berks RG20 0JD

The American Boxwood Society
PO Box 85, Boyce, VA 22620, USA

Lyn Batdorf, Curator of Box,
U S National Arboretum,
3501 New York Avenue NE,
Washington D.C. 20002-1958 USA

PUBLICATIONS

Box Hill, The National Trust, 1997.
The guide book gives further information about Box Hill.

The Boxwood Bulletin published by the American Boxwood Society contains information on cultivars and other general topics.

Box in English Place Names by Richard Coates will be included in a forthcoming volume of *English Studies.*

There is very little published information on the ecology of box, but most floras give a basic description.

Batdorf, Lyn R, *Boxwood Handbook.* American Boxwood Society, 1994. This practical guide to growing box is obtainable from the RHS shop at Wisley or from the Langley Boxwood Nursery.

Strong R, *Small Traditional Gardens.* Conran Octopus, London, 1992. Gives historical information on gardens, with plenty of references to box.

Whalley, R and Jennings, A, *Knot Gardens and Parterres: A history of the knot garden and how to make one.* Barn Elms, 1998

There are various books on topiary; these two give historical background and are practical guides. Although out of print, they can be obtained through libraries.

Clevely A M, *Topiary, The Art of Clipping Trees and Ornamental Hedges.* Collins, 1988.

Hadfield, Miles, *Topiary and Ornamental Hedges.* London, 1971.